Frozen Secrets: The Untold Mysteries Beneath Antarctica

Aurora K. Winter

Published by Antillas Publishing House, 2025.

FROZEN SECRETS: THE UNTOLD MYSTERIES BENEATH ANTARCTICA

First edition. January 18, 2025.

ISBN: 979-8230921738

Written by Aurora K. Winter.

Table of Contents

About the Author: Aurora K. Winter

Aurora K. Winter is an environmental writer and explorer with a profound passion for Earth's most extreme landscapes. Combining a background in ecological studies with a lifelong fascination for polar regions, she delves into the mysteries of ice-covered frontiers to share their significance and wonder with the world. *Frozen Secrets: The Untold Mysteries Beneath Antarctica* is a testament to her dedication, blending adventure and research into an engaging journey across the last great wilderness on our planet.

Introduction: The Allure of the Unknown

In the far reaches of the Southern Hemisphere lies a realm so stark, so silent, and so untouched that it feels like another planet. **Antarctica**—the world's last uncharted frontier—rises from the ocean in a sprawling expanse of ice and rock, majestic and intimidating. Here, you'll find no ancient ruins built by human hands, no bustling cities, and no roads that snake across its frozen plains. Instead, you encounter an unbroken wilderness, swept by howling katabatic winds and cloaked in perpetual whiteness.

I first became fascinated by Antarctica during my university years, when I stumbled upon a documentary showing satellite imagery of the continent. I remember gazing at those swirling, white vortexes of ice, riveted by the fact that so much of our planet still lay beyond the footprints of modern civilization. Something about the sheer emptiness and mystery of it captured me then—and never let go. Over time, as I read about the heroic age of exploration—legends like Roald Amundsen, Robert Scott, and Ernest Shackleton—I realized that Antarctica was more than just an ice desert. It was a testament to human courage, curiosity, and tenacity.

But the deeper I delved, the more I recognized that Antarctica is not simply about heroic tales; it is also about **untapped secrets** locked beneath miles of ice. It's a vault of knowledge that could potentially reshape our understanding of **climate change**, **astrobiology**, and **Earth's geological past**. Hidden lakes beneath the ice sheet might harbor life forms that have evolved in complete isolation for millennia. Meteorites that have fallen onto this frozen landscape remain remarkably preserved, holding clues to the origins of our solar system and even the possibility of life elsewhere.

This book is my invitation for you to join me on a journey into this icy realm. We will relive some of the most daring expeditions, examine the scientific endeavors unraveling the planet's climate history, and dive into the enigma of subglacial lakes and hidden ecosystems. We will also touch upon the myriad ways in which Antarctica influences global weather patterns, ocean currents, and sea levels—connecting this isolated continent to every corner of the Earth.

Above all, my hope is that through these pages, you will come to see Antarctica not as a barren wasteland, but as a vibrant and mysterious world in its own right. It is a land that demands respect and caution, yet offers boundless wonder to those brave enough to explore its depths. Welcome, dear reader, to **Frozen Secrets: The Untold Mysteries Beneath Antarctica**.

Chapter 1: A Brief History of Antarctic Exploration

1.1 The First Dreamers and The Age of Myths

Long before any human being set foot on Antarctica, ancient philosophers and scholars speculated about a vast southern landmass—"Terra Australis Incognita." They imagined it as a continent balancing the Earth in the south, much as Eurasia balanced it in the north. Maps from the 15th and 16th centuries sometimes depicted a mythical southern continent. This concept captured the imagination of cartographers, who drew elaborate shorelines even though no solid evidence supported their claims.

The *real* story of Antarctic discovery begins with maritime explorers who ventured into uncharted southern latitudes. Early navigators, such as **James Cook**, sailed across the southern seas in the late 18th century, but formidable ice fields thwarted their progress. Cook famously proclaimed that if there was a southern continent, no man could approach it given the perilous conditions. His assertion only fueled curiosity. If an entire land lay hidden beneath the ice, why was it so inaccessible?

1.2 Cracking the Ice: Early Landings and Races

By the 19th century, whale and seal hunters began to push further into the Southern Ocean, encountering icebergs larger than anything previously recorded. Several teams claimed to have spotted the Antarctic coastline. In 1820, three separate expeditions (led by Fabian von Bellingshausen, Edward Bransfield, and Nathaniel Palmer) arguably sighted continental Antarctica around the same time. Yet, these sightings were fleeting and none resulted in a documented landing on the mainland.

The first individual credited with stepping onto the Antarctic continent is often cited as **John Davis**, an American sealer, in 1821. However, historical records are murky, and it remains a topic of debate among polar historians. Even so, these early explorers encountered unimaginable challenges: colliding sea ice, unpredictable storms, and a lack of reliable navigation charts. The harshness of the environment demanded extraordinary bravery and resilience.

1.3 The Heroic Age: Scott, Amundsen, and Shackleton

The **Heroic Age of Antarctic Exploration** (roughly 1895 to 1922) was a defining era that introduced the world to legendary figures who risked everything to uncover the continent's secrets. Technologies of the time—wooden ships reinforced with steel, primitive motors, and nascent radio communication—were hardly sufficient for such an extreme environment. Yet, with an unbreakable spirit, these explorers pressed on.

- **Robert Falcon Scott (United Kingdom)**: A Royal Navy officer who led two expeditions (the Discovery Expedition from 1901 to 1904 and the Terra Nova Expedition from 1910 to 1913). Scott's meticulous scientific data collection was groundbreaking, even if his push for the South Pole ended tragically.
- **Roald Amundsen (Norway)**: A master of polar travel who prioritized practicality, dog sleds, and survival

techniques learned from indigenous Arctic peoples. Amundsen's team became the first to reach the South Pole on December 14, 1911. His victory was overshadowed by Scott's later demise.

- **Ernest Shackleton (United Kingdom)**: Celebrated for his leadership and the *Endurance* expedition (1914–1917), during which his ship was crushed by ice. Shackleton famously led his men to safety against impossible odds, a testament to extraordinary fortitude.

These expeditions, fraught with danger, romance, and tragedy, enthralled the public. They also illuminated the scientific significance of Antarctica. Despite heartbreak—such as Scott and his companions perishing on the ice—heroic tales of perseverance reshaped humanity's view of the seventh continent. Antarctica was no longer a myth; it was real, tangible, and formidable.

1.4 Twentieth-Century Progress and the Birth of Science in Antarctica

After the Heroic Age, technological advances and global interest in polar science led to new expeditions with more specialized goals. Planes gradually replaced dogsleds. Diesel-powered icebreakers and radio communication improved logistics. Explorers like **Richard E. Byrd** of the United States pioneered the use of aircraft in mapping. Byrd's multiple expeditions from the 1920s to the 1950s offered unprecedented aerial surveys of inland Antarctica.

During World War II, several nations established Antarctic bases, albeit with varied motivations. Some aimed for territorial claims; others saw strategic advantages. This shift laid the groundwork for the post-war collaboration that would define modern Antarctica. Scientific stations sprouted across the continent, enabling experts to conduct research in geology, meteorology, oceanography, and glaciology.

In 1959, the **Antarctic Treaty** was signed, designating Antarctica as a scientific preserve and banning military activity. This treaty was an extraordinary act of global diplomacy, setting a precedent for peace and cooperation in a region that could otherwise have sparked conflict. From that moment on, Antarctica became synonymous with collaboration and cutting-edge research.

1.5 Lessons and Legacy

The early explorers exemplify the powerful lure of the unknown and the incredible lengths humans will go to satisfy curiosity. Their diaries and logs revealed that Antarctica is a place of **unforgiving extremes**—minus 50 degrees Celsius is not uncommon, with winds capable of toppling well-prepared explorers. Yet the lessons they gleaned have informed modern polar travel: the importance of thorough preparation, appropriate clothing and gear, and respect for the environment's dangers.

This heroic chapter in Antarctic history also set the stage for the **advanced scientific inquiries** that define current research. Those battered wooden ships and dog teams paved the way for the satellite-based observations and subglacial drilling projects we see today. It is in the spirit of those early pioneers that modern-day scientists continue to push the boundaries of knowledge, revealing the hidden wonders beneath the ice.

Chapter 2: The Continent of Ice—Geography and Climate Extremes

2.1 The Silent White Wilderness

Picture this: an endless expanse of white in every direction, meeting a brilliant blue sky at the horizon. The daylight is intense, magnified by the whiteness beneath. On the coldest days, the air is utterly still, and any sound you do make—footsteps crunching over snow, breathing through frozen scarf layers—feels magnified. Antarctica is often described as a **"polar desert"**, an environment that receives less precipitation than the Sahara. Yet, paradoxically, it holds the largest volume of ice on Earth.

Covering nearly **14 million square kilometers** (though this fluctuates seasonally with sea ice), Antarctica dwarfs most continents when you consider its ice shelves. But the image of a flat, homogenous ice sheet is misleading. Beneath that ice, you'll find mountain ranges—like the Gamburtsev Mountains—that rival the Alps in scale, hidden from view. The weight of thousands of meters of ice exerts such immense pressure on the continent that large parts of Antarctica's bedrock lie below sea level. This interplay between mountainous terrain and thick ice is one reason exploring the continent is such a challenge.

2.2 Glacial Giants and Shifting Ice Shelves

Antarctica's ice is not static. Glaciers flow outward, and at the coasts, these rivers of ice fan out into **floating ice shelves**. Some of the more famous ones include the **Ross Ice Shelf** and the **Ronne-Filchner Ice Shelf**, each spanning an area larger than many countries. Ice shelves form a critical buffer, slowing the flow of land-based ice into the ocean. When they weaken or collapse, glacial ice can more rapidly slip seaward, contributing to rising sea levels.

- **Ice Streams**: These are swift-flowing corridors of ice within the ice sheet, analogous to rivers in a desert. They can be tens of kilometers wide, moving meters to kilometers per year.
- **Crevasses**: Deep cracks that form as the ice flows over uneven terrain or accelerates. They present significant hazards to explorers and scientists on the ground.

Despite what appears to be a frozen, motionless environment, Antarctic ice is in a **constant state of slow evolution**. Satellite observations allow scientists to track the rate at which glaciers and ice shelves advance or retreat, providing invaluable data for global climate models.

2.3 The Climate That Redefines "Extreme"

Antarctica is the **coldest, windiest, and driest** continent on Earth. In some inland regions, the average annual temperature can be as low as **-57°C** (-70°F). The lowest natural temperature recorded on Earth, **-89.2°C** (-128.6°F), was measured at Vostok Station in East Antarctica. Surviving in this environment demands specialized gear, technology, and constant vigilance.

Katabatic Winds

One of Antarctica's most notorious meteorological phenomena is the **katabatic wind**. This occurs when cold, dense air accumulates over the elevated ice sheet and then rushes downhill toward the coast under the force of gravity. These winds can reach hurricane speeds, blasting anything in their path with bone-chilling gusts. Explorers have recounted hearing the roar of these winds before actually feeling them, a disconcerting audio prelude to an onslaught of piercing cold.

Polar Day and Polar Night

Because of Antarctica's position at the Earth's southern pole, it experiences extreme variations in daylight throughout the year. During the **austral summer** (roughly October to February), the sun barely sets, and "midnight sun" reigns. This constant daylight can be psychologically disorienting, making regular sleep schedules challenging. Conversely, the **austral winter** plunges the continent into months of near-total darkness, with only brief periods of twilight. Overwintering scientists must adapt to months without direct sunlight, relying on artificial light to maintain a semblance of normalcy.

2.4 Locked Archives of Earth's Past

Perhaps the most scientifically significant aspect of Antarctica's ice is that it serves as a **time capsule** of Earth's climate history. As snow accumulates, layer by layer, it traps air bubbles and microscopic particles. Over centuries, this snow compresses into firn and eventually solid ice, each layer preserving a snapshot of the atmosphere at the time it was formed.

Researchers drill **ice cores** to retrieve these layers, analyzing trapped gases—especially carbon dioxide and methane—to reconstruct **paleoclimates**. By correlating variations in greenhouse gas concentrations with temperature proxies from isotopes of hydrogen or oxygen, scientists can piece together climate changes over **hundreds of thousands of years**. Ice cores from Antarctica have helped us understand natural climate oscillations, such as glacial and interglacial periods, and how current human-driven changes compare to geological benchmarks.

2.5 Pushing Human Limits

All of this makes Antarctica not only a harsh environment for wildlife but also a proving ground for human endurance and ingenuity. Settlements here aren't "permanent" in the traditional sense; rather, they're **research stations** staffed by rotating teams of scientists, support personnel, and adventurers. Life in these outposts revolves around ensuring basic needs—heat, shelter, food, and safety—while pushing scientific boundaries.

- **Logistical Challenges**: Transporting equipment and supplies requires meticulous planning. Ships and icebreakers deliver heavy cargo during the short summer season, while planes handle personnel transfers on specialized runways made of ice.
- **Medical Preparedness**: Each station typically has at least one trained doctor, and telemedicine consultations with teams in other parts of the world

are common. Evacuation in winter can be impossible, so medical self-sufficiency is a must.

Yet, for those who venture there, the reward is an experience like no other. Crisp, clear nights reveal star-filled skies with an intensity rarely seen elsewhere on Earth. Rare optical phenomena such as **auroras** and **moon halos** become routine, transforming the environment into an ever-changing spectacle of light and shadow.

2.6 The Larger Ties That Bind

Antarctica's climate does not exist in isolation. The intense cold and persistent winds help drive **global circulation patterns**. Cold water sinking around Antarctica influences ocean currents worldwide, affecting marine ecosystems as far north as the equator. In turn, phenomena like El Niño and La Niña can alter weather patterns in the Southern Ocean, demonstrating the planet's interconnectedness.

Understanding Antarctica's climate is crucial to predicting **global sea-level rise**. Because so much of the world's freshwater is locked in Antarctic ice, even small increases in melting can translate into significant changes for coastal communities around the globe. Models must account for how warming ocean waters nibble at the edges of ice shelves from below and how shifting weather patterns alter precipitation on the ice sheet.

A Compass for Climate Science

In many ways, Antarctica acts as a **compass for climate science**. Rising temperatures and changing ocean currents often manifest in the polar regions first, serving as an early warning for the rest of the planet. Within the pages ahead, we will dive deeper into these topics, discovering how the seemingly silent continent echoes with messages crucial for our shared future.

Chapter 3: Life at the Edge — Creatures and Adaptations in Antarctica

3.1 Adaptations to a Lethal Cold

Biting winds that can peel flesh, salt-laden blizzards whipping across the ice, and temperatures so low they can freeze human skin in seconds—this is the reality faced by Antarctic organisms. Yet, life here does not merely survive; it **thrives** in ingenious ways. When you picture the wildlife of Antarctica, you might imagine **Emperor Penguins** gracefully marching across the ice or robust **Weddell seals** lounging on snow drifts. These animals represent only a fraction of the continent's resilient inhabitants.

Penguins: Icons of Polar Resilience

- **Emperor Penguins** (Aptenodytes forsteri): Standing nearly four feet tall, Emperor Penguins endure one of the harshest breeding cycles on the planet. As the sea ice expands in winter, males incubate the precious single egg on their feet, huddling together in colossal groups to share body heat. Without access to food for months, they subsist on their fat reserves until the female returns, having fed in the open ocean.
- **Adélie Penguins** (Pygoscelis adeliae): Smaller but no less robust, Adélies rely on strong social bonds for defense and warmth. Their comical "tobogganing"—sliding on their bellies across the

ice—helps them conserve energy while searching for open water leads rich in krill.

Behind the penguins' appeal lies a suite of **physiological marvels**: extra layers of blubber, tightly packed feathers that trap air for insulation, and a specialized circulatory system that minimizes heat loss in their extremities.

Marine Mammals: Masters of the Chill

- **Weddell Seals**: Famous for their big eyes and calm demeanor, these seals can dive more than 600 meters (2,000 feet) in search of fish and squid. They keep breathing holes open by grinding their teeth against the ice, ensuring they can surface for air even when the sea freezes over.
- **Leopard Seals**: Sleek and formidable hunters, leopard seals prowl along the ice edges, ambushing penguins and smaller seals. Despite their fierceness, they're a critical component of Antarctica's ecological puzzle, preventing overpopulation of other species and maintaining balance.

Each marine mammal has evolved a **dense layer of blubber**—an insulating and energy-storing tissue—that helps maintain internal temperatures when the surrounding water can be well below zero (saltwater, after all, can remain liquid at sub-freezing temperatures).

Winged Invaders in the Sky

Not all Antarctic birds are flightless. **Skuas** and **petrels** roam these skies, sometimes traveling thousands of kilometers from nesting sites to find food. The **snow petrel**, pure white against the backdrop of ice, is a stunning sight. Like most Antarctic birds, they time their breeding to coincide with the brief burst of summer productivity, when krill and fish abound near the surface.

3.2 The Miracle of Microbes: Life in Dark, Icy Corners

While penguins and seals get the limelight, some of the most astonishing Antarctic inhabitants are invisible to the naked eye. **Microbes—bacteria, archaea, fungi, and single-celled algae—**hide within sea ice, soil pockets, and even below the ice sheet in subglacial lakes. These creatures e one by exploiting every possible ecological niche.

Extremophiles: Thriving Where Others Perish

- **Psychrophiles**: Microbes that not only tolerate but **require** cold temperatures to live. They produce specialized enzymes and antifreeze proteins that keep their cellular structures stable at sub-zero conditions.
- **Halophiles**: Found in high-salinity brine channels within sea ice, these organisms have adapted to both cold and high salt levels, which would be lethal to most other lifeforms.

For a long time, scientists believed Antarctica's inland regions were too harsh for sustained microbial activity. However, **drilling projects** in places like the Dry Valleys—a region that receives almost no precipitation—have unearthed microbial communities in the permafrost. Their ability to remain dormant for extended periods and then "revive" when conditions improve fascinates astrobiologists. If life can survive here, perhaps it can survive on Mars or the icy moons of Jupiter and Saturn.

Biological Inspiration

The proteins and enzymes that allow Antarctic microbes to function could have significant commercial and scientific applications. Enzymes extracted from psychrophiles are often used in laboratory processes that require lower temperatures—reducing energy costs and preserving delicate substrates. Antarctica is, in many ways, Earth's grand laboratory for understanding **extreme biology**.

3.3 Lessons for Humanity

Antarctica's species embody the principle of **adaptation**: they alter their physiology, behavior, and social structures to endure extreme cold, isolation, and scarce resources. In a broader sense, these biological strategies resonate with our own struggles in a rapidly changing world.

1. **Resourcefulness**: Emperor Penguins and Weddell seals show that energy conservation is crucial in environments of scarcity.
2. **Collaboration**: Penguin huddles illustrate the power

of community efforts to combat external threats (in their case, deadly cold).
3. **Resilience**: Microbes surviving under Antarctic ice for millennia highlight life's capacity to persist under conditions once deemed impossible.

As we grapple with **climate change**, diminishing resources, and political uncertainties, Antarctica's creatures remind us that survival often hinges on adaptability and cooperation. Like the Emperor Penguins, humanity may find strength in unity; like the microbes, we may need to embrace creativity and flexibility to sustain life in harsh conditions.

With this appreciation for Antarctic life, we move forward to a hidden realm beneath the ice: the subglacial lakes. These buried bodies of water might hold entire ecosystems waiting to be discovered, offering the next frontier of exploration in a place we once thought empty and lifeless.

Chapter 4: Frozen Subworlds — The Secrets of Subglacial Lakes

4.1 Lake Vostok: A World Apart

Imagine a lake, larger than Lake Ontario, sealed under nearly four kilometers of ice for **millions of years**. That is **Lake Vostok**, one of the largest and deepest subglacial lakes in Antarctica. It was first suspected to exist in the mid-20th century, but only through satellite radar and seismic measurements in the later decades did scientists confirm its presence.

- **Isolation**: What makes Lake Vostok so compelling is the possibility of ancient, isolated life forms. Cut off from the surface, sunlight, and any external influence, the ecosystem—if it exists—would have had to adapt in near-total darkness and high pressure.
- **Warmth and Circulation**: Surprisingly, subglacial lakes can maintain liquid water due to geothermal heat from Earth's interior and the immense pressure exerted by the ice overhead. In some cases, the friction generated by moving ice also contributes to warming. Scientists believe Lake Vostok has under-ice currents that circulate nutrients, hinting at the potential for a stable environment.

Drilling into Lake Vostok has been fraught with **scientific and ethical challenges**. There are genuine concerns about contaminating a pristine habitat. Scientists must use sterile techniques, specialized lubricants, and closed drilling systems to ensure that whatever life may dwell there remains untainted by modern Earth microbes.

4.2 The Mysterious Depths of Other Subglacial Lakes

Lake Vostok, though the most famous, is only one among **hundreds** of subglacial lakes mapped beneath Antarctica. Some are relatively small, mere pockets of meltwater, while others link through networks of subglacial rivers.

Hidden Networks Beneath the Ice

- **Lake Whillans**: Located beneath the Whillans Ice Stream, this lake periodically drains and refills, demonstrating dynamic water flow beneath the ice sheet. Researchers have drilled here more recently to collect water samples, discovering microbial life that may feed on iron and minerals from the bedrock.
- **Ellsworth Subglacial Lake**: Nestled under the West Antarctic Ice Sheet, Lake Ellsworth has drawn interest due to its depth and potential for unique microorganisms. An ambitious drilling project took place in 2012, though technical difficulties meant it didn't penetrate the lake as planned.

These hidden lakes vary widely in depth, salinity, and chemistry. The patterns of flooding and draining also affect how ice above them moves. Observing those patterns via satellites helps scientists understand how water flows under glaciers, and thus how quickly large sections of the ice sheet might respond to climate shifts.

Technological Frontiers

Exploring subglacial environments requires specialized tools:

- **Hot-water drilling** systems to melt through kilometers of ice without introducing contaminants.
- **Remote submersibles** or *ROVs (Remotely Operated Vehicles)* adapted to crushing pressures and low temperatures.
- **Deep ice core** sampling, where each layer of ice reveals a partial story of past climate.

4.3 Windows to Other Planets

One of the most exciting implications of subglacial lake research is its relevance to **astrobiology**. Icy moons such as **Europa** (orbiting Jupiter) and **Enceladus** (orbiting Saturn) are believed to harbor subsurface oceans beneath their frozen crusts. If life can exist in the dark, pressurized depths of Antarctic lakes, it strengthens the possibility of life existing on these distant worlds.

- **Europa's Ocean**: Evidence suggests a global ocean beneath Europa's thick ice, warmed by tidal forces. Salts and other compounds on the surface hint at a potentially chemical-rich sea below—an environment not so different from a subglacial Antarctic lake.

- **Enceladus**: Geysers of water vapor and organic molecules erupt from the moon's south pole, indicating liquid reservoirs beneath the ice. Studying Antarctica's subglacial lakes provides a terrestrial analog for how microorganisms might survive on Enceladus.

These parallels drive **NASA** and other space agencies to fund Antarctic research as a testing ground for instruments and drilling methods that might one day penetrate the ice of alien oceans. Indeed, some of the robots designed to explore Lake Vostok or Lake Whillans could serve as prototypes for interplanetary missions.

Transition to the Next Chapters

As we've now seen, Antarctica is far more than an icy desert. It is a **hive of evolutionary wonders**, from the largest penguins on Earth to the tiniest microbes beneath layers of ice. And it gets stranger still: **meteorites** that land here remain astonishingly well-preserved, hinting at cosmic secrets. **Stations** of scientific endeavor dot the landscape, hosting humans for months on end in conditions that test both body and mind. The fate of these ice shelves and glaciers also teeters under a changing climate, raising global stakes.

In the following chapters, we will explore:

- How **meteorites** discovered in Antarctica provide insights into our solar system's birth.
- The daily lives of **scientists and workers** who maintain research stations in perpetual cold and darkness.

- The growing role of **technology** in unraveling mysteries that even the first explorers never dreamed existed.
- The urgent message Antarctica sends about **climate change** and the future of our planet.

Stay with me as we continue to unearth Antarctica's **frozen secrets**—secrets that span from the smallest cells under the ice to interplanetary frontiers. The world's coldest continent may indeed be our hottest key to understanding life, Earth's past, and perhaps even the cosmic tapestry itself.

Chapter 5: Messages from the Cosmos — Meteorites in the Antarctic Ice

5.1 A Gift from the Heavens

High on the Antarctic plateau, amid sweeping winds and biting cold, scientists walk slowly, eyes fixed on the icy surface. They are searching for **meteorites**—fragments of asteroids, comets, and even other planets that have fallen through Earth's atmosphere. Antarctica is uniquely suited for meteorite hunting. The bright white ice acts like a natural display case, contrasting sharply with the dark stones. Over centuries, the continent's glacial movements and wind patterns have concentrated these extraterrestrial treasures in certain areas, making them easier to find here than almost anywhere else on Earth.

The value of these meteorites extends beyond their scarcity or exotic allure. Each fragment tells a story of the **solar system's early formation**. Some samples predate Earth itself, carrying chemical fingerprints of ancient cosmic processes. Others, known as **Martian meteorites**, could offer clues to the Red Planet's geology—and possibly, its habitability. In these small, dark stones, entire chapters of cosmic history await unraveling.

5.2 Why Antarctica?

One might wonder why meteorite expeditions often choose such a forbidding environment. The reasons are as clear as Antarctic ice:

1. **Visibility**: Dark meteorite fragments stand out against the snow and ice, making them relatively easy to spot.
2. **Preservation**: Low temperatures and arid conditions help preserve meteorites by minimizing weathering and chemical reactions.
3. **Glacial Concentration**: Over time, ice flow can trap and funnel meteorites toward "stranding surfaces" or blue ice fields, where wind erosion exposes them on the surface.

Ironically, this very harshness also poses immense challenges. Field teams must brave fierce katabatic winds, sub-zero temperatures, and the risk of hidden crevasses as they traverse the ice. Their perseverance, however, pays off: each new find can potentially rewrite a page in our cosmic history books.

5.3 ANSMET and Other Expeditions

A prime example of these organized efforts is the **ANSMET** program (Antarctic Search for Meteorites), initiated in the late 1970s. Funded largely by NASA and the U.S. National Science Foundation, the program sends small teams of researchers to remote sites for weeks at a time. Living in tents pitched on ice, these explorers use snowmobiles to crisscross carefully selected regions.

Fieldwork Adventures

- **Identifying Candidates**: Researchers scan the horizon for dark objects. When one is spotted, the team stops to inspect its shape, texture, and color. A meteorite often appears smoother on one side due to partial melting during atmospheric entry.

- **Sampling**: If confirmed to be a meteorite, the stone is photographed and placed in a Teflon or polyethylene bag. This simple step is crucial for preventing contamination—Earth-based microbes or chemicals could alter the specimen's authenticity.
- **Logistics**: Once the field season ends, meteorites are sent to labs (often in the United States or elsewhere) for classification and detailed study. Many labs have special clean rooms and curation facilities to handle these precious materials safely.

Though ANSMET is one of the most famous programs, other international teams conduct parallel searches. Collaborations with Japanese, European, and Chinese Antarctic programs further broaden the scope, ensuring a constant stream of meteorite discoveries.

5.4 Decoding the Past: What Meteorites Reveal

Each meteorite is like a tiny time capsule, preserving:

- **Chemical Signatures**: Ratios of elements (like iron, nickel, magnesium) point to an origin in specific parts of the solar system.
- **Isotopic Clues**: Radioactive decay products yield an age estimate. These measurements have confirmed that some meteorites formed around 4.5 billion years ago—roughly the same age as the solar system.

- **Planetary Origins**: Certain compositions match the chemical makeup of Mars or the Moon, proving that cosmic collisions can eject planetary debris into space.

Martian Meteorites

Occasionally, a stone from Antarctica is discovered to be of **Martian origin**. When Mars was younger and likely more volcanically active, asteroid impacts blasted rocks off its surface and into space. Some eventually fell to Earth. These samples are invaluable for **astrobiology**: if they hold signs of fossilized microbial life or particular mineral signatures (like carbonates), we could gain insights into early Martian environments. While no definitive proof of ancient life has been found yet, the ongoing research keeps the hope alive.

5.5 Guardian of Secrets: Antarctica's Role in Astrobiology

Antarctica's role in preserving cosmic artifacts ties neatly with the idea of **life on other worlds**. If extremophiles under Antarctic ice can survive with minimal nutrients, what of similar creatures on, say, Europa or Enceladus? Likewise, meteorites from the distant reaches of the solar system remind us that we are part of a larger cosmic network. Each meteorite is a messenger, bridging the vast expanse between planets and offering glimpses into the universe's early chapters.

And while the hunt for meteorites has a tinge of romantic adventure—gliding across a polar desert in search of relics from space—it is also rigorous science. Each discovery supports ongoing missions to Mars, helps calibrate instruments used on rovers, and enriches our understanding of planet formation and the possibility of life elsewhere. Antarctica, in this respect, becomes not just a terrestrial frontier, but a cosmic crossroads.

Chapter 6: Bases of Discovery — Humanity's Foothold on the Ice

6.1 Shelters at the End of the World

Scattered across the Antarctic ice, clusters of buildings cling to the whiteness like outposts on a distant planet. These are **research stations**—the lifelines of modern polar science. Some, like **McMurdo Station** (run by the United States) or **Rothera Research Station** (operated by the British Antarctic Survey), resemble tiny polar towns with multiple labs, dormitories, and communal halls. Others, such as smaller field camps, might consist of just a few huts and tents erected seasonally.

Life here is precarious yet purposeful. Every aspect of survival demands diligence: an errant spark could set a station ablaze, water sources must be meticulously managed, and communication lines must remain open in an environment where help is a thousand kilometers away.

6.2 Living in Perpetual Winter (and Summer)

The human experience on the ice is shaped by extremes of light and dark:

- **Summer**: From October to February, days stretch endlessly. The sun circles overhead, never quite dipping below the horizon. Aircraft can land more frequently, ship deliveries become possible, and research teams can make extended forays into the field. People often experience sleep disturbances due to the continuous daylight; black-out curtains and sleep masks become essential.

- **Winter**: From March to September, the continent is locked in near-complete darkness. Only a skeleton crew remains, often fewer than 50 people at major stations. They maintain critical experiments, ensure station upkeep, and support the occasional rescue or re-supply flight (though these are risky in the winter night). Morale can waver, and psychological support is crucial.

Despite these challenges, a strong sense of camaraderie typically develops among station inhabitants. Shared meals, movie nights, or festive events on holidays anchor a semblance of normal life in this otherworldly context. For many, the closeness and shared purpose turn Antarctica into a second home, albeit one perpetually on the brink of a winter storm.

6.3 The Architecture of Survival

Constructing and maintaining buildings on the coldest continent requires engineering marvels:

1. **Insulation**: Walls, floors, and roofs are heavily insulated to minimize heat loss. Double-door "cold porches" help reduce icy gusts when people enter or leave.
2. **Elevated Structures**: Many newer stations are built on platforms or stilts, allowing wind to blow underneath. This prevents snow accumulation around the buildings that could eventually bury them.

3. **Renewable Energy**: Some stations experiment with **wind turbines** and **solar arrays** to reduce reliance on diesel generators. However, harsh conditions and limited sunlight in winter make year-round renewable energy a technical challenge.

Station managers also plan for worst-case scenarios. Emergency shelters, backup generators, and carefully stocked pantries can sustain a small community for months if supply routes close.

6.4 Daily Rhythms: From Science to Survival

Daily life in an Antarctic research station intertwines routine tasks with extraordinary science. A glaciologist might spend half the day calibrating sensors measuring ice movement, then switch to shoveling snow off solar panels. Everyone takes turns in kitchen duty, cleaning communal areas, and managing waste responsibly, as environmental regulations are stringent. Water, for instance, is often recycled and rationed. Showers might be limited to a couple of minutes a few times per week.

Psychological Landscape

The environment can feel isolating, particularly in winter when contact with the outside world is limited to email and occasional satellite calls. Close quarters test interpersonal dynamics. Small disagreements can loom large when you see the same faces each day and know there is no immediate escape.

To counter these stresses, stations often encourage group activities: communal dinners, scientific lectures, film nights, talent shows—even improvised music jam sessions. The sense of shared mission fosters resilience. When storms rage and the temperature dips below -40°C, the knowledge that everyone relies on each other cements a strong bond.

6.5 Human Impact in a Pristine Land

Even with strict environmental protocols, human presence leaves a footprint. Waste disposal systems must be carefully managed to prevent pollution. Research vessels near the coast must navigate carefully to avoid disrupting marine life. Tourism, albeit tightly regulated, has grown in recent decades, raising concerns about large cruise ships venturing into delicate ecosystems.

The Antarctic Treaty System

Since 1959, the **Antarctic Treaty** has governed this continent's use. Key tenets include:

- **Demilitarization**: No military bases or weapons testing.
- **Scientific Freedom**: Unimpeded research and data sharing among nations.
- **Environmental Protection**: Protocols on waste disposal, wildlife conservation, and limiting human impact.

Additional agreements, such as the **Protocol on Environmental Protection to the Antarctic Treaty (Madrid Protocol)**, further safeguard Antarctica's ecosystem. Still, climate change, marine resource exploitation (like krill fishing), and increasing tourist traffic pose **ongoing challenges**. Researchers at the stations become frontline witnesses to these evolving pressures.

6.6 On the Frontier of Knowledge

At its core, each station's purpose is to further scientific discovery. Geologists map ancient rock formations, climatologists monitor shifting wind patterns, biologists study extreme life forms, and astronomers exploit the high altitude and clear skies to look into the depths of the universe. Many stations house sophisticated labs, satellite uplinks, and sensor networks streaming real-time data to collaborators across the globe. Antarctica has even hosted cutting-edge detectors for neutrino research (like the **IceCube Neutrino Observatory** at the South Pole), leveraging the continent's vast ice as a sensor medium.

In many ways, each station serves as a **microscopic city of the future**—self-contained, cooperative, and resource-conscious. Lessons learned in Antarctica about sustainability, conflict resolution, and scientific collaboration may prove invaluable for humanity's survival on Earth—and someday, for living on other planets.

Chapter 7: Technology and the Future — Robots, Satellites, and the Next Frontier

7.1 Robots on Ice: Pushing Beyond Human Limits

Imagine a small, torpedo-shaped vehicle diving beneath the Antarctic ice shelf, navigating total darkness and bone-chilling temperatures to map a subglacial cavity. No human diver could survive there, but a **remotely operated vehicle (ROV)** or an **autonomous underwater vehicle (AUV)** can. Robots like these have transformed Antarctica from an "explorer's nightmare" into a "technologist's playground," offering unprecedented access to places humans could never reach safely.

AUVs and ROVs: Eyes Under the Ice

- **Data Gathering**: High-resolution cameras, sonar, LiDAR, and chemical sensors collect imagery and readings of water chemistry, under-ice topography, and biological activity.
- **Navigation Challenges**: Under-ice missions require advanced navigation; GPS signals do not penetrate thick ice, so these vehicles rely on internal inertial guidance systems or transponder networks deployed beneath the ice shelf.
- **Sample Retrieval**: Some ROVs are equipped with mechanical arms that can gently scoop sediment from

the seafloor or scrape biofilms from ice surfaces, enabling direct analysis of sub-ice ecosystems.

Stories of robotic missions often read like polar adventures of their own. Researchers describe moments of elation when a vehicle resurfaces intact and laden with data—and moments of despair when an ROV is lost in an under-ice cavern. Each success or failure helps refine new generations of polar-capable robots, inching us closer to a comprehensive picture of Antarctica's hidden realms.

7.2 Satellites Over the White Continent

Where robots conquer the under-ice darkness, **satellites** survey Antarctica's grandeur from space. Earth-orbiting instruments have revolutionized our ability to study the vast ice sheets, track shifting glaciers, and monitor environmental changes in near-real time.

- **Radar Altimetry**: By sending radar pulses from a satellite and measuring their echo, scientists can determine the height of the ice surface down to a matter of centimeters. This helps detect even minute changes in ice thickness or glacial flow.
- **Gravity Field Mapping**: Missions like **GRACE (Gravity Recovery and Climate Experiment)** have detected variations in Earth's gravity field that indicate mass loss in major Antarctic ice streams.
- **Optical and Infrared Imaging**: By studying differences in surface reflectivity, temperature, and

color, satellites can distinguish between ice, water, and rock, as well as track seasonal changes in sea ice coverage.

Such remote sensing tools **unveil patterns** invisible to ground-level observers. For instance, a glacier front can calve icebergs that look static and unremarkable up close. But from orbit, an ice shelf may be retreating at a rate of kilometers per year—a clear marker of changing climate conditions. Satellites have also discovered previously unknown subglacial lakes, adding to Antarctica's tapestry of hidden aquatic networks.

7.3 The Role of Drones and Aerial Surveys

Drones (UAVs) fill the gap between ground-based studies and satellite observations. They can hover closer to the ice surface, capturing ultra-detailed images while covering large areas. In the polar environment, drone pilots must contend with fierce winds, sub-zero temperatures that sap battery life, and the challenge of landing on an ever-shifting ice field.

- **Mapping Crevasse Fields**: Drones armed with LiDAR can create 3D maps of treacherous crevasse zones, improving safety for field parties.
- **Wildlife Monitoring**: Biologists use drones to observe penguin colonies or seal populations from above, minimizing disturbance to sensitive habitats.
- **Real-Time Crisis Response**: When an ice shelf shows signs of imminent calving or collapse, drones can quickly gather data to assess potential hazards.

7.4 High-Tech Field Stations

Antarctic stations, especially newer or refurbished ones, increasingly resemble **cutting-edge research hubs**. Many feature advanced computing clusters linked via satellite communications. Continuous streams of telemetry from weather stations, seismic instruments, and oceanographic buoys feed into a global data network.

Engineers also experiment with **green technologies**—wind turbines, solar panels, and battery systems—to reduce the reliance on diesel generators. The challenge of operating in near-perpetual darkness during winter remains, but each success story paves the way for more sustainable polar science.

7.5 Antarctica as a Testbed for Space Exploration

Because Antarctica shares traits with extraterrestrial environments—extreme cold, isolation, and harsh weather—tech designed for polar conditions often has **spinoff applications** for space missions. NASA and other space agencies regularly test rovers, drilling systems, and life-support technologies in Antarctica. If a rover can handle the pressure under an Antarctic ice shelf, perhaps it can do the same on **Europa**. If a habitat can maintain warmth and power through a polar winter, that knowledge might inform strategies for lunar or Martian bases.

Such synergies highlight Antarctica's **dual identity**: it is both an otherworldly realm on our own planet and a stepping stone to further cosmic exploration. In short, the southernmost continent is not just a subject of scientific inquiry, but a crucible where tomorrow's engineering feats are forged.

Chapter 8: Relics of the Past, Keys to the Future — Ancient Climate Records

8.1 The Ice Core Chronicles

Antarctica's ice is more than frozen water; it is a **time capsule** storing Earth's climate secrets. Each snowfall layer compresses into firn and eventually into solid ice, trapping tiny bubbles of ancient atmosphere. By drilling deep into the ice sheet and carefully extracting long cylinders of ice—known as **ice cores**—scientists can measure past concentrations of carbon dioxide (CO_2), methane, and other greenhouse gases.

Some of the longest ice cores have reached back **800,000 years** into Earth's past, revealing **eight glacial-interglacial cycles**. These cyclical changes in temperature and atmospheric gas levels shed light on how Earth's climate fluctuates naturally—and how our current era of rapidly rising CO_2 fits into that grand timeline.

Reading the Layers

- **Bubble Analysis**: Each trapped air bubble is a tiny relic of prehistoric atmosphere, allowing direct comparison of ancient greenhouse gas levels with today's.
- **Isotopic Composition**: Ratios of oxygen isotopes in the ice reflect temperature shifts over time. By analyzing variations in these isotopes, scientists can

reconstruct past temperature records with remarkable precision.

- **Dust and Pollutants**: Layers of volcanic ash, pollen, or industrial pollutants reveal major events—eruptions, large-scale fires, or anthropogenic activities.

8.2 The Stories Told by Ancient Sediments

Beyond ice cores, Antarctica's hidden geological archives also offer clues about ancient warm periods and ice-free intervals. On the continent's periphery or along submarine ridges, scientists gather **sediment cores** from beneath the seafloor. These sediments can contain microscopic fossils of marine organisms (like diatoms) that thrived in warmer waters, indicating times when the Antarctic coast was much less icy.

Astonishingly, **fossilized remains** of forests discovered in certain mountainous regions suggest that Antarctica once hosted lush vegetation—an era long before ice gripped the continent. By dating these fossils and correlating them with global tectonic and climatic events, researchers piece together a narrative of radical environmental shifts. Antarctica has been many things over geological time: a piece of Gondwana, a green forested expanse, and ultimately, the ice-crowned landmass we see today.

8.3 Climate Lessons for the Modern World

Why do these ancient records matter? Because they offer **context**. By showing how Earth's climate system responded to natural forces over hundreds of thousands—or even millions—of years, these records provide a benchmark against which we can measure our current situation.

- **Historical CO_2 Levels**: During past warm spells, atmospheric CO_2 levels approached 300 parts per million (ppm). Today, we are well over 400 ppm—and climbing.
- **Sea-Level Rise**: Sediment and ice core records show that when global temperatures rose, polar ice sheets shrank, raising sea levels significantly. This underscores the vulnerability of modern coastlines.

Breaking the Feedback Loops

One crucial insight from ice core data is the interplay between **CO_2** and **global temperature**. Warming oceans release more CO_2, which amplifies warming further, creating a feedback loop. Conversely, cooling oceans absorb more CO_2, promoting further cooling. These loops, once triggered, can accelerate climate shifts. Modern fossil fuel emissions risk pushing these feedbacks in a direction that leads to rapid, large-scale changes.

8.4 Watching the Shifts in Real Time

Today's glaciologists don't just rely on ice core data and geologic evidence; they can **observe** changes happening as we speak. Satellite altimetry shows decreasing ice mass in parts of West Antarctica. Tide gauge records and satellite measurements confirm that sea levels are on the rise. Climate models, calibrated with data from Antarctica's past, project potentially dramatic changes if greenhouse gas emissions remain unchecked.

Paleoclimatology and Policymaking

Paleoclimatologists often present their findings at international forums, influencing **global climate policy**. When negotiators hear that Earth's polar ice has historically collapsed in conditions not much warmer than our own, they gain a tangible perspective on future risks. Antarctica, in essence, becomes a **voice of caution**—one that echoes over eons of Earth's history, urging us to tread carefully in the decades to come.

8.5 Glacial Wake-Up Call

The deeper we drill into Antarctic ice, the clearer our planet's cyclical nature appears. Yet this cyclical pattern is now colliding with an accelerated, human-driven shift. Antarctica's ancient records stand as a stark warning: just because changes occurred gradually in the past doesn't guarantee a gentle trajectory forward. If enough tipping points are crossed—such as the destabilization of major ice shelves—sea levels could surge, displacing hundreds of millions of people worldwide.

Still, the story is not solely one of doom. These very same records also reveal **resilience**. Life bounced back after mass extinctions, ice retreated and advanced, and Earth found new equilibria. The question is whether humanity can manage the transition without catastrophic disruption, learning from the largest climate archive on the planet.

Chapter 9: Reliving the Extreme — Modern Adventurers on the Ice

9.1 Rekindling the Spirit of Discovery

Ever since Roald Amundsen first planted a Norwegian flag at the South Pole and Robert Falcon Scott's team tragically followed, Antarctica has beckoned those with an unquenchable thirst for challenge. Today, the era of thick wooden ships and dog teams has largely passed, but modern adventurers carry the same fire in their hearts. They journey not just for scientific progress but to test the **limits of human endurance**, forging new legends in the world's harshest terrain.

From individuals attempting solo, unsupported ski crossings to international teams vying to summit unclimbed peaks, these contemporary explorers echo the feats of the Heroic Age. Yet they are aided by advanced gear—lightweight tents built to withstand hurricane-force winds, GPS tracking devices, and hyper-efficient stoves that melt snow into drinking water. While technology increases safety, the core challenge remains: Antarctica's storms, extreme temperatures, and psychological isolation can vanquish even the most prepared traveler.

9.2 The Modern Solo Expeditions

One of the most captivating trends has been the rise of **solo expeditions**—adventurers who traverse the ice entirely alone, dragging sleds laden with food and supplies for weeks or months at a time. Their routes may follow historical paths or carve entirely new trails across unexplored ridges and ice fields.

- **Mapping Their Way**: High-resolution satellite images guide these explorers, but storms can quickly obscure landmarks and GPS can fail in extreme cold. Many adventurers rely on old-fashioned compasses or sun navigation as a backup.
- **Psychological Battles**: Loneliness poses one of the greatest hurdles. When the nearest human is hundreds of kilometers away, a missed satellite phone call or a broken piece of gear can feel catastrophic. Journaling, voice-recording diaries, and scheduled check-ins with support teams help stave off despair.
- **Diet and Caloric Demands**: Pulling a sled across snow requires an immense energy output—easily surpassing 5,000 calories per day. Dehydrated meals, energy bars, and high-fat items like cheese, nuts, and chocolate fill expedition rations.

While some claim records for speed or distance, most solo trekkers emphasize personal growth. Surviving a polar crossing, they attest, is as much an internal transformation as a physical one—an exercise in *resilience, grit, and acceptance* of one's insignificance in the face of nature's vast power.

9.3 Scaling Antarctic Peaks

Antarctica's ice plateau is commonly perceived as flat and monotonous, but **mountain ranges** do break its expanse: the **Transantarctic Mountains**, the spectacular **Ellsworth Mountains**, and lesser-known peaks cloaked in perpetual ice. Alpinists find these ranges extremely challenging:

- **Vinson Massif**: At 4,892 meters (16,050 ft), this is the highest peak on the continent. Climbing Vinson is logistically complex, requiring aircraft that can land on remote, icy runways and base camps that endure bitter winds.
- **Uncharted Summits**: Many smaller Antarctic peaks remain unnamed and unclimbed, attracting elite mountaineers who long for the thrill of true exploration. Confronting hidden crevasses and fast-changing weather, they must balance ambition with safety at every step.

9.4 Adventure Tourism: Boon or Bane?

While solo expeditions and elite climbs garner headlines, another trend has emerged—**adventure tourism**. Each year, a growing number of "extreme tourists" seek to check Antarctica off their bucket lists, whether by camping near the South Pole or skiing part of the way. Specialized tour operators promise "the ultimate trip," complete with guided visits to penguin colonies, short helicopter flights, or even an overnight stay on the ice.

Though heavily regulated by the International Association of Antarctica Tour Operators (IAATO), this influx of tourists raises concerns:

- **Environmental Impact**: More foot traffic and potential pollution threaten fragile habitats.
- **Safety**: Tourists often lack polar experience, relying on guides for survival.
- **Ethical Questions**: Should Antarctica, protected as a

common heritage of mankind, become another checkmark for thrill-seekers?

Proponents argue that well-managed tourism fosters global appreciation and stewardship for the continent, turning visitors into advocates. Opponents warn that even minimal disturbance can degrade pristine ecosystems. Balancing these viewpoints remains an ongoing dialogue among scientists, policymakers, and the local tourism sector.

9.5 The Call of the Ice

Despite the debates, Antarctica retains its allure. It is a place to **rediscover our own fragility**, to challenge personal boundaries, and to stand in awe of an environment that remains largely untouched by humankind. From novice adventurers to hardened explorers, each returns with stories of ephemeral beauty—jagged ice cliffs illuminated by slanting sunlight, roaring winds that sculpt snow into eerie shapes, and the humbling sensation of being utterly alone on a continent that dwarfs human ambition.

In many ways, these modern odysseys echo the deeper reasons humanity has always sought out wild places: not simply for conquest, but for a renewed understanding of **ourselves**. Antarctica, with its punishing conditions and breathtaking horizons, offers a mirror to reflect our greatest weaknesses and highest aspirations.

Chapter 10: Perils and Promises — Antarctica in a Changing Climate

10.1 The Front Line of Global Warming

As much as Antarctica's pristine landscapes symbolize nature's grandeur, they also stand at the **front line of climate change**. Rising atmospheric temperatures and shifting ocean currents threaten to reshape the continent's ice sheets—and, in turn, the world's coastlines. Scientists warn that Antarctica's massive reserves of freshwater ice could dramatically raise global sea levels if key ice shelves weaken and collapse.

Warming Trends in West Antarctica

West Antarctica, in particular, is vulnerable due to its **marine-based ice sheet**, where the ice rests on bedrock below sea level. Warming ocean waters can invade beneath ice shelves, thinning them from below. Once an ice shelf loses its integrity, glaciers feeding it can accelerate, sliding more ice into the sea—a phenomenon known as **marine ice sheet instability**. Places like **Thwaites Glacier**, ominously nicknamed the "Doomsday Glacier," epitomize these fears.

10.2 The Carbon Connection

Why does a spike in global CO_2 levels matter so much for Antarctica? The Earth's climate system is interconnected in numerous, sometimes **surprising** ways:

1. **Ocean Circulation**: Warmer surface waters from lower latitudes can travel south, melting ice shelves. Changes in salinity and temperature also affect deep-ocean currents, which, in turn, regulate global heat distribution.
2. **Feedback Loops**: As ice melts, less sunlight is reflected back into space (due to the lower albedo of exposed rock or open ocean), leading to further warming.
3. **Atmospheric Changes**: Shifts in wind patterns, linked to global warming, can affect how sea ice forms around Antarctica or steer warm air over typically frigid regions.

Though the impacts may seem distant, these processes can ripple outward, influencing weather patterns and sea levels far beyond the polar regions.

10.3 The Southern Ocean Under Siege

Encircling Antarctica is the **Southern Ocean**, a critical cog in Earth's climate machinery. It absorbs a significant portion of the planet's excess heat and a substantial fraction of anthropogenic CO_2 emissions—effectively buffering the rest of the world from more extreme climate swings. Yet this buffering capacity is not infinite. As the ocean warms and becomes more acidic, it jeopardizes marine life adapted to cold, stable conditions.

- **Krill Populations**: Antarctic krill form the foundation of the Southern Ocean food web, feeding penguins, seals, and whales. They depend on sea ice for part of their life cycle. If ice coverage or ocean acidity levels shift too drastically, krill numbers could plummet.

- **Ocean Acidification**: The Southern Ocean is absorbing CO_2, which dissolves to form carbonic acid, reducing the pH of seawater. Shell-forming organisms like pteropods may struggle to maintain their shells, destabilizing an entire food web.

10.4 Melting at the Margins: Ice Shelf Collapse Events

The **Larsen Ice Shelf** system on the Antarctic Peninsula provides a stark example of what can happen when warming takes hold. In 1995, **Larsen A** disintegrated; in 2002, **Larsen B** followed suit. Within mere weeks, a chunk of ice roughly the size of Rhode Island shattered into thousands of icebergs. The collapse shocked scientists, who had expected such a dramatic event to unfold over decades, not days.

Subsequent research revealed that meltwater pooling on the ice shelf surface likely contributed to fracturing. As climate models predict further warming, other ice shelves—especially in West Antarctica—face similar risks.

10.5 A Tipping Point or a Curve We Can Bend?

The term **"tipping point"** often arises in discussions about Antarctic ice. It refers to a threshold beyond which changes become self-sustaining and hard to reverse. Once a major glacier or ice shelf passes such a threshold, the resulting ice discharge can continue even if temperatures later stabilize. This raises pressing questions about our global carbon trajectory: how quickly might we approach these tipping points, and is there a path to avoid them?

International Collaboration

Antarctica's future is not just the domain of scientists and polar explorers. **Diplomats, policymakers, and global organizations** play a pivotal role in mitigating climate change. The **Antarctic Treaty System** remains one of the most successful examples of international cooperation. Could this spirit of collaboration expand to address carbon emissions, renewable energy, and environmental protection worldwide?

Technological Solutions

Renewable energy, carbon capture and storage, and improved efficiency in industries may help slow emissions. However, even the best-case scenarios project **significant warming** in polar regions. Adaptation strategies—from coastal defenses for human populations to more stringent protection of Antarctic marine environments—may become necessary. Yet the cost of adaptation could be massive if we fail to stabilize global emissions.

10.6 Hope Amid the Ice

While the data can be sobering, Antarctica also inspires **hope**. The continent's story over geological time is one of resilience and transformation. Life there continues to surprise us with its adaptability—from microbes in subglacial lakes to penguins finding new feeding grounds when ice shelves collapse. Humanity, too, has proven capable of innovation and collaboration against the odds.

Scientists argue that **actions taken now**—policy shifts, energy transition, and conservation efforts—can significantly alter the trajectory of polar ice loss over the coming century. The fate of Antarctica, in many ways, reflects the fate of the entire planet. By listening to the warnings emanating from the frozen south, we might yet steer a course that preserves its majestic ice sheets and the global stability they safeguard.

Chapter 11: Myths, Legends, and Conspiracy Theories

11.1 The Allure of the Unknown

Since the first mapmakers marked "Terra Australis Incognita" at the bottom of their charts, Antarctica has captured the human imagination. In modern times, this fascination takes many forms—some rooted in genuine scientific curiosity, others spun from pure speculation. For an icy land so remote and inhospitable, Antarctica has spawned an abundance of **myths, legends, and conspiracy theories.** From ancient lost civilizations beneath the ice to secret military installations, rumor and reality often intertwine in the polar night of the world's most enigmatic continent.

11.2 Ancient Civilizations and the Piri Reis Map

One longstanding myth claims that Antarctica was once home to an **advanced ancient civilization.** Proponents point to the **Piri Reis map**, a 16th-century chart drawn by the Ottoman admiral Piri Reis. The map intriguingly appears to depict parts of Antarctica's coastline free of ice. According to some interpretations, this suggests knowledge of the continent before it was officially discovered and even before it was frozen—leading to wild theories about lost races or ancient mariners with advanced cartographic skills.

- **Scholarly Explanations**: Historians believe the map is more likely to show parts of South America or an imaginative extension of its coastline. The "Antarctic" sections may be misinterpretations or artistic flourishes.
- **Geological Evidence**: Scientific consensus places Antarctica under ice for millions of years, rendering any advanced human civilization there extraordinarily unlikely.

11.3 UFOs, Hollow Earth, and Alien Bases

A quick online search reveals countless claims that Antarctica hides alien bases or functions as an entry point to a **"hollow Earth."** Some stories reference bizarre aerial anomalies recorded by polar explorers, while others weave in actual historical footnotes—such as the intense interest in polar regions by various militaries—into elaborate tales of **extraterrestrial contact**.

- **Operation Highjump (1946–47)**: A real U.S. Navy operation led by Admiral Richard E. Byrd to establish a research presence in Antarctica. Conspiracy theorists have rebranded it as a secret war against hidden Nazi bases or extraterrestrial forces.
- **Scientific Rebuttals**: There is no credible evidence for alien activity under the ice. Radar, seismic surveys, and drilling data reveal a world of subglacial mountains, lakes, and geological strata—not alien tunnels or

hollow caverns leading to an inner Earth.

11.4 Nazi Bases and World War II Legends

Another persistent legend suggests that the Nazis established secret bases in Antarctica during World War II. Tales of "Neuschwabenland" recount how German expeditions discovered ice-free oases and built underground fortresses, from which they supposedly planned a post-war resurgence.

While there **was** a German Antarctic Expedition in 1938–39—which named part of the continent "Neu-Schwabenland" and dropped pennants onto the ice from an aircraft—there is no documented evidence of any permanent German base or advanced fortress. Historical records show their activities were mostly exploratory: surveying and mapping the coastline. The dire logistics and minimal strategic value make it highly implausible that hidden Nazi installations ever existed.

11.5 Creatures of Legend: Cryptids and Sea Monsters

Antarctica's seas teem with life—penguins, seals, whales, and a thriving ecosystem of krill. Yet rumors of **sea monsters** or **cryptids** occasionally appear in anecdotal reports. Some explorers have spun yarns about colossal "ice serpents," while conspiracy websites claim the existence of giant octopuses in subglacial lakes.

- **Plausible Origins**: Marine animals can appear monstrous when glimpsed under unusual conditions or from a distance. Large squid species do inhabit deep

Antarctic waters, fueling tall tales.

- **Scientific Perspective**: No credible evidence exists for unknown "sea monsters." The real wonders—like colossal squid or unknown extremophiles—are fascinating enough without resorting to fiction.

11.6 The Power of Imagination

Why does Antarctica inspire so many fanciful stories? Its **remoteness** and **harshness** make it an ideal blank canvas onto which we project our fears and fantasies. Even in an age of satellites and scientific stations, much of Antarctica remains physically inaccessible. This natural barrier to exploration nurtures the notion that **something hidden** could be out there—whether a lost city, advanced beings, or secret technologies.

Ultimately, these legends, while mostly debunked by science, add to the continent's cultural tapestry. They demonstrate the enduring human desire to believe in the extraordinary and the unknown. Antarctica, in that sense, remains our planet's final frontier—a place vast enough to accommodate both rigorous science and boundless imagination.

Chapter 12: Epilogue — The Frozen Legacy

12.1 A Glimpse Back at the Journey

As we conclude our exploration of **Antarctica**, it is both humbling and exhilarating to reflect on the ground we have covered. We began with the first audacious explorers, who risked life and limb to reach this ice-clad continent. We delved into Antarctica's **geography**, **biology**, and **hidden subglacial realms**. We examined **meteorites** that connect us to distant corners of the solar system, learned about **life in research stations**, and surveyed the advanced **technology** that allows us deeper insights into this unforgiving environment.

We also confronted the stark realities of a **warming climate**, observing how sea ice decline and melting glaciers link Antarctica's fate with our own. Finally, we ventured into the realm of **legends**, acknowledging the power of myth in shaping our perception of the unknown.

12.2 Antarctica's Role as a Global Symbol

From its beginnings as a blank spot on medieval maps to its modern status as a hub of international scientific cooperation, Antarctica has been a **symbol** of humanity's broader relationship with Earth. It is a place where:

- **Politics can step aside**: The Antarctic Treaty and subsequent protocols demonstrate that nations can indeed set aside competing interests for the sake of shared exploration and conservation.

- **Climate change is magnified**: Antarctic research underscores global interconnectedness, revealing how shifts at the poles can spell changes for coastal cities worldwide.
- **Innovation thrives**: Extreme conditions breed cutting-edge solutions—from renewable energy experiments at research stations to sophisticated robotics probing subglacial lakes.

12.3 Protecting the Last Frontier

While Antarctica's ecosystems are resilient in their own right—penguins, seals, and microscopic extremophiles endure unimaginable cold—human activity poses significant challenges. Overfishing, potential mineral exploitation, irresponsible tourism, and climate-driven ice loss threaten the delicate balance. Continued vigilance and **international cooperation** are needed to safeguard this unique continent.

- **Marine Protected Areas (MPAs)**: Efforts to establish large MPAs around Antarctica's waters can help ensure the safety of krill populations and the entire food web that depends on them.
- **Strict Environmental Protocols**: Ongoing enforcement of regulations under the Antarctic Treaty System is vital to minimize pollution and habitat disturbance, particularly as tourism grows.

12.4 The Future of Discovery

Despite decades of research, **much remains unknown** about Antarctica. New technologies—autonomous sub-ice vehicles, better satellite imaging, more capable drilling rigs—promise breakthroughs. We may yet discover microbial life beneath the ice that could hold medical or astrobiological significance. Meteorite searches could unearth the next big clue about our solar system's infancy. And as we refine our climate models, Antarctica's ice cores will remain a cornerstone of understanding Earth's past.

For students and aspiring scientists, the continent stands as a **living laboratory**—a place to tackle fundamental questions in geology, biology, glaciology, and beyond. For adventurers, it remains an arena to test one's mettle against nature's harshest elements. For dreamers, it is a canvas for boundless imagination.

12.5 An Invitation to Wonder

Perhaps the greatest lesson Antarctica offers is one of **wonder**. Its stark, unyielding beauty strips away the superfluous, leaving only raw landscapes and elemental forces—wind, ice, sea, and sky. Here, more than anywhere, one feels the planet's heartbeat. The calls of penguins carry across vast plains of ice; the roar of calving glaciers echoes like thunder; the auroras dance overhead in colors too vivid for words.

In this era of rapid change, Antarctica is a reminder of Earth's fragility—an epic stage on which the drama of global climate unfolds in real time. If we heed its warnings and celebrate its wonders, we may find both the humility and the resolve to safeguard our collective future. Antarctica stands as a silent guardian of our past, a mirror reflecting our present, and a harbinger of what awaits us if we fail to act.

With that final reflection, we close this book—but the story, of course, does not end here. Antarctica awaits the next generation of explorers, scientists, and visionaries, holding its secrets under ice and sky. May it continue to inspire awe and unity among all who hear the call of the **Frozen South**.

This concludes our journey through

Frozen Secrets: The Untold Mysteries Beneath Antarctica

by *Aurora K. Winter*.

Author's Note (Aurora K. Winter):

Thank you for accompanying me to the end of the world—and perhaps, the beginning of a greater understanding of our planet. May Antarctica's silent expanses and hidden wonders fuel your curiosity and commitment to protect this astonishing realm.

www.ingramcontent.com/pod-product-compliance
Lightning Source LLC
LaVergne TN
LVHW022315200926
848481LV00004B/1357
9798230921738